ASTROPHYSICS FOR NOOBS

Contents

Introduction to Astrophysics
    What is Astrophysics?
    The Universe and Its Components
    The Tools of the Trade
    A Brief History
    Astrophysics vs. Astronomy vs. Cosmology
    Modern Astrophysics
    Why It Matters

The Scale of the Universe
    Tiny to Huge: The Universe's Size
    Really Small Things: Quantum World
    Everyday Size: Human Scale
    Our Cosmic Neighbourhood: The Solar System
    Our Star City: The Milky Way Galaxy
    Neighbourhood of Galaxies: The Local Group
    The Part We Can See: The Observable Universe
    Beyond Our Sight: The Entire Universe
    Why This Matters

Theories of the Universe
    The Big Bang Theory: The Universe's Fiery Start
    Steady State Theory: The Unchanging Universe
    The Oscillating Universe Theory: The Cosmic Bounce
    The Multiverse Theory: Many Universes
    String Theory: The Universe's Tiny Strings

The Ekpyrotic Universe: The Universe's Fiery Collision
Loop Quantum Gravity: The Universe's Fabric
The Holographic Principle: The Universe's Projection
Why These Theories Matter

Observable Universe vs. Unknown Universe
The Observable Universe: Our Cosmic Visibility
The Speed of Light: The Cosmic Speed Limit
The Unknown Universe: Beyond Our Sight
The Universe's Expansion: Getting Bigger All the Time
Why It Matters: The Big Cosmic Picture
The Future of Observation: New Horizons

The Structure of the Sun
The Sun's Layers: A Cosmic Onion
Nuclear Fusion: The Sun's Powerhouse
Why It's Fascinating
The Big Picture

Main Sequence Stars
What Are Main Sequence Stars?
The Life Cycle of a Main Sequence Star
The Diversity of Main Sequence Stars
The Importance of Mass
Our Sun: A Typical Main Sequence Star
Energy Production: Nuclear Fusion
The End of the Main Sequence

Why Main Sequence Stars Matter

Giants and Supernovas
Supernova: The Universe's Fireworks
The Role of Nuclear Fusion in Type Ia Supernova
What makes Type Ia Supernova Special
The Mystery Behind Type Ia Supernova

The Solar System
The Sun: The Heart of the Solar System
The Planets: A Diverse Family
Dwarf Planets and Other Residents
Moons: Planets' Companions
The Asteroid and Kuiper Belts
The Edge of the Solar System
Why the Solar System Matters
Exploration and Discovery

Terrestrial vs. Gas Giants
Terrestrial Planets: The Rocky Worlds
Gas Giants: The Massive Atmospheres
Key Differences
Why the Distinction Matters
Exploring Beyond Our Solar System

Moons
Moons: The Travelling Companions of Planets
Natural Satellites: Born, Not Made
The Role of Gravity

The Importance of Moons
Moons as Targets for Exploration
The Moon: Our Closest Neighbour

Black Holes
Black Holes: The Cosmic Enigmas
Formation of Black Holes
Detecting Black Holes
The Mystery and Importance of Black Holes
The Future of Black Hole Research
Stellar-Mass Black Holes: The Star's End
Supermassive Black Holes: Galactic Giants
Intermediate-Mass Black Holes: The Missing Link
Primordial Black Holes: Echoes of the Big Bang
Why Different Types Matter
The Ongoing Mystery

Wormholes
Wormholes: The Universe's Tunnels
Visualising Wormholes
Types of Wormholes
The Challenges
The Role of Wormholes in Science Fiction
The Future of Wormhole Research

Types of Galaxies
Spiral Galaxies: The Cosmic Pinwheels
Elliptical Galaxies: The Star-Studded Spheres
Irregular Galaxies: The Unconventional Outliers

Why Galaxy Types Matter
Exploring the Universe

Galaxy Clusters
Galaxy Clusters: The Universe's Great Cities
The Intracluster Medium: The Space Between
Gravitational Giants
Galaxy Groups vs. Clusters
Superclusters: Clusters of Clusters
Studying Galaxy Clusters
The Role of Dark Matter

Dark Matter
Dark Matter: The Invisible Substance
Evidence for Dark Matter
What Could Dark Matter Be?
Searching for Dark Matter
The Mystery Continues
The Accelerating Universe: A Speeding Cosmos
The Relationship Between Dark Energy and Dark Matter
Unravelling the Mysteries
Why Dark Matter Matters

Exoplanets
Understanding Exoplanets
Methods of Detection
Types of Exoplanets
Habitable Zones and Life

The Future of Exoplanet Research
Why Exoplanets Fascinate Us

Introduction to Astrophysics

What is Astrophysics?

Astrophysics is a fascinating and expansive field that sits at the crossroads of physics and astronomy. It's all about applying the principles of physics and chemistry to unravel the mysteries of the universe. Let's break it down into more digestible parts:

The Universe and Its Components

At its core, astrophysics seeks to understand the universe and everything in it. This includes studying the birth, life and death of stars; the intricate workings of galaxies; the mysterious nebulae and even the planets - including our own. Astrophysicists look at these celestial objects and phenomena to get a better grasp of the universe's grand design.

The Tools of the Trade

Astrophysicists use a variety of methods from physics to study space. They rely on observing the light and other forms of radiation that come from space to learn about the temperature, composition and movement of celestial bodies. They also use theoretical models to predict how these objects will behave and interact.

A Brief History

The field of astrophysics is relatively young. It began to take shape with the work of Sir Isaac Newton, who showed that the same physical laws that govern motion on Earth also apply to the heavens. This was a revolutionary idea because it meant that the universe was not governed by different sets of rules for celestial and terrestrial objects.

Astrophysics vs. Astronomy vs. Cosmology

Astrophysics often gets mixed up with astronomy and cosmology, but there are subtle differences:

- **Astronomy** is the oldest of these sciences and is more about measuring and cataloguing celestial objects—like mapping the stars and noting their movements.
- **Astrophysics** goes a step further by trying to understand the physical properties and processes of these objects—like figuring out what stars are made of or how they form and evolve.
- **Cosmology** zooms out even further to consider the universe as a whole, tackling questions about its origin, structure and ultimate fate.

Modern Astrophysics

Today, astrophysics is a dynamic field that covers a lot of ground. Astrophysicists might study how stars are born from clouds of dust and gas, or they might investigate the nature of dark matter and dark energy—mysterious substances that make up most of the universe but are invisible to us. They also search for planets around other stars, looking for signs of life beyond Earth.

Why It Matters

Astrophysics helps us understand where we come from and our place in the cosmos. It tells us about the elements that make up our bodies and the earth beneath our feet. It's a quest for knowledge that has practical benefits too, like the development of new technologies and materials.

In summary, astrophysics is the science that uses the laws of physics to make sense of the universe. It's a journey of discovery that has taken us from wondering about the twinkling lights in the night sky to exploring the possibility of life on distant worlds.

The Scale of the Universe

Tiny to Huge: The Universe's Size

The universe is like a giant puzzle with pieces that range from super small to unbelievably big. Let's start from the smallest bits and work our way up to the really huge stuff.

Really Small Things: Quantum World

At the very bottom, we have things that are so small you can't see them, not even with a microscope. These are called subatomic particles and they're the tiny bits that make up everything.

Everyday Size: Human Scale

Moving up, we reach the size of things we see around us every day, like people, animals, cars and buildings. This is the human scale and it's where we live our lives.

Our Cosmic Neighbourhood: The Solar System

If we zoom out a bit, we find our solar system. It's our home in space, with planets like Earth that go around a big, bright star we call the Sun. The distances here are so big that light, which is super fast, takes about 8 minutes to travel from the Sun to us.

Our Star City: The Milky Way Galaxy

Keep zooming out and you'll see our galaxy, the Milky Way. It's like a giant city of stars, with billions of them, including our Sun. If you could travel at the speed of light, it would take you 100,000 years to get from one side to the other!

Neighbourhood of Galaxies: The Local Group

Our Milky Way is just one of many galaxies that hang out together in a space neighbourhood called the Local Group. It's a collection of more than 50 galaxies and it's pretty spacious.

The Part We Can See: The Observable Universe

When we talk about the observable universe, we mean the part of space we can actually see from Earth. It's a bubble around us that's about 93 billion light-years across. A light-year is the distance light travels in a year, which is really, really far—about 9.46 trillion kilometres!

Beyond Our Sight: The Entire Universe

There's more to the universe than what we can see. In fact, it might be endless. We just don't know because we can't see that far, not yet anyway.

Why This Matters

Understanding how big the universe is helps us figure out our place in it. It's amazing to think about how big space is and how small we are in comparison. It also makes us curious and inspires us to learn more.

Theories of the Universe

The Big Bang Theory: The Universe's Fiery Start

The most popular theory is the Big Bang Theory. It suggests that the universe started as a tiny, super-hot and dense point about 13.8 billion years ago. Then, it expanded and cooled down to form everything we see today. Think of it like a tiny seed that suddenly grew into a vast cosmic garden.

Steady State Theory: The Unchanging Universe

An alternative to the Big Bang is the Steady State Theory. This idea says the universe looks the same no matter where or when you observe it. It suggests that new matter is constantly created as the universe expands, keeping its appearance unchanged over time. Imagine a bakery that keeps making bread so there's always the same amount on the shelves, even as people take loaves away.

The Oscillating Universe Theory: The Cosmic Bounce

This theory proposes that the universe goes through cycles of big bangs and big crunches, expanding and contracting like a giant cosmic lung. Each cycle is like a universe's lifetime, from birth to death, before starting all over again.

The Multiverse Theory: Many Universes

The Multiverse Theory is like saying our universe is just one bubble in a giant cosmic foam bath. There could be countless other universes out there, each with its own laws of physics and properties.

String Theory: The Universe's Tiny Strings

String Theory suggests that the universe's fundamental particles aren't points, but tiny, vibrating strings. These strings could be the key to combining gravity with quantum mechanics, which are the rules for the smallest things in the universe. It's like finding the common thread that weaves everything together.

The Ekpyrotic Universe: The Universe's Fiery Collision

The Ekpyrotic Theory comes from the Greek word for 'conflagration' and suggests that our universe is the result of a collision between two three-dimensional worlds on a hidden fourth dimension. It's like two sheets of paper touching in a space we can't see, creating a spark that starts a new universe.

Loop Quantum Gravity: The Universe's Fabric

Loop Quantum Gravity tries to describe the fabric of space and time using loops of quantum threads. It's a bit like a fishnet, with each knot representing a tiny bit of space-time, showing that the universe isn't smooth but has a structure at the smallest scales.

The Holographic Principle: The Universe's Projection

This principle suggests that all the information in the universe can be stored on a two-dimensional surface, like a hologram. It's as if everything we see in three dimensions is actually a projection from a flat, two-dimensional space.

Why These Theories Matter

Understanding different theories of the universe helps us explore possibilities beyond what we currently know. It's like having different maps to treasure; each one might lead us to new discoveries about our cosmic home.

Observable Universe vs. Unknown Universe

The Observable Universe: Our Cosmic Visibility

The Observable Universe is like our cosmic backyard. It's the part of the universe that we can see or could see in theory, using telescopes and other tools. It's a bubble around Earth, extending out to about 93 billion light-years. This bubble is not fixed; it grows as time passes because light from farther away has more time to reach us.

The Speed of Light: The Cosmic Speed Limit

The speed of light is super important here. It's the fastest anything can travel in the universe. Because of this limit, we can only see things whose light has had enough time to get to us since the universe began. That's why we talk about the "observable" part—it's all about the light's journey through space and time.

The Unknown Universe: Beyond Our Sight

Now, the Unknown Universe is like the parts of an ocean we haven't explored. It's everything that exists beyond what we can observe. We think the universe might be infinite, with no edges or borders. But we can't see it all because the light from those parts hasn't reached us yet, or maybe it never will.

The Universe's Expansion: Getting Bigger All the Time

One of the coolest things about the universe is that it's expanding. This means that the observable universe is getting bigger as space itself stretches out. So, objects that were once close enough for their light to reach us are now moving away. It's like friends walking away from you; the farther they go, the harder it is to see them.

Why It Matters: The Big Cosmic Picture

Understanding the difference between the observable and unknown universe helps us grasp the limits of our knowledge. It's a reminder that there's so much more out there to discover. It also tells us about the history of the universe because looking far away means looking back in time.

The Future of Observation: New Horizons

As technology gets better, we'll be able to see more of the universe. New telescopes and instruments will let us peek further into our cosmic backyard and maybe even catch a glimpse of the unknown parts. It's an exciting time to be curious about the universe!

In summary, the observable universe is the part we can see, while the unknown universe is everything else that's out there, waiting to be discovered. It's a vast, mysterious and incredibly exciting frontier that challenges us to keep exploring and learning.

The Structure of the Sun

The Sun's Layers: A Cosmic Onion

The Sun is like a giant onion with different layers, each playing a crucial role in its structure and function.

- **Core**: At the very centre is the core, where the temperature and pressure are so high that hydrogen atoms fuse together to form helium. This process is called nuclear fusion and it's what makes the Sun shine.
- **Radiative Zone**: Surrounding the core is the radiative zone. Here, energy produced in the core moves outward very slowly, taking thousands of years to pass through.
- **Convection Zone**: Next is the convection zone, where hot gases rise, cool and then sink back down. This movement helps transport energy toward the Sun's surface.
- **Photosphere**: This is the visible surface of the Sun. It's where the light we see is emitted and it's also where sunspots, which are cooler, darker areas, can be found.
- **Chromosphere**: Above the photosphere is the chromosphere, a layer of the Sun's atmosphere that glows with a reddish colour during solar eclipses.
- **Corona**: The outermost layer is the corona, which is much hotter than the layers below and

extends millions of kilometres into space. It's usually seen during a total solar eclipse as a white halo.

Nuclear Fusion: The Sun's Powerhouse

Nuclear fusion is the heart of the Sun's energy production. It's a process where the Sun's immense gravity forces hydrogen atoms to come together and form helium, releasing a huge amount of energy.

- **Proton-Proton Chain**: This is the main fusion process in the Sun. It starts with two hydrogen nuclei (protons) merging to form deuterium, a type of heavy hydrogen. In a series of steps, more protons join and helium is produced, along with energy in the form of light and heat.
- **Energy Journey**: The energy from fusion first travels through the radiative zone, then the convection zone and finally, it reaches the photosphere. From there, it radiates out into space as sunlight.

Why It's Fascinating

The Sun's structure and the process of nuclear fusion are not just fascinating; they're essential for life on Earth. The energy produced by the Sun heats our planet and provides the light needed for plants to grow.

The Big Picture

Understanding the Sun's structure and how nuclear fusion works gives us insight into the workings of other stars and the vast universe. It's a reminder of the incredible forces of nature that operate on a cosmic scale.

In summary, the Sun is a complex, layered star powered by nuclear fusion. This process is the source of all the heat and light we receive from the Sun, making it the ultimate energy source for our planet.

Main Sequence Stars

What Are Main Sequence Stars?

Main sequence stars are the powerhouse of the universe. They are stars that are in the prime of their life, fusing hydrogen into helium in their cores. This process releases a tremendous amount of energy, which we see as light and feel as heat from our Sun.

The Life Cycle of a Main Sequence Star

A star begins its life as a cloud of dust and gas. Under the force of gravity, this cloud collapses and as it does, it heats up. Once the core gets hot and dense enough, nuclear fusion starts. This is the process that powers the star and makes it shine. The star has now entered the main sequence phase of its life.

The Diversity of Main Sequence Stars

Main sequence stars come in a variety of sizes and colours, from small, cool, red dwarfs to massive, hot, blue giants. Their mass can range from about a tenth of the mass of the Sun to up to 200 times as massive.

The Importance of Mass

The mass of a main sequence star determines its brightness, colour and how long it will live. Larger stars burn brighter and hotter but have shorter lives because they use up their fuel more quickly. Smaller stars, like red dwarfs, burn their fuel slowly and can live for tens to hundreds of billions of years—much longer than the current age of the universe.

Our Sun: A Typical Main Sequence Star

Our Sun is a typical G-type yellow dwarf star. It's been shining for about 4.6 billion years and has about 5 billion years to go before it runs out of hydrogen in its core and leaves the main sequence.

Energy Production: Nuclear Fusion

The core of a main sequence star is where the magic happens. Here, hydrogen atoms are fused together to form helium in a process that releases energy. This energy travels out from the core and eventually reaches the surface of the star, making it shine.

- **Proton-Proton Chain**: For stars like the Sun and smaller, the fusion process is called the proton-proton chain. It's a series of reactions that build helium from hydrogen and release energy.
- **CNO Cycle**: For larger stars, the CNO cycle is the dominant fusion process. It uses carbon, nitrogen and oxygen as catalysts to help fuse hydrogen into helium.

The End of the Main Sequence

A star remains on the main sequence as long as it has hydrogen to fuse in its core. When the hydrogen runs out, the star leaves the main sequence. What happens next depends on the star's mass. It might expand into a red giant, or if it's massive enough, it could end its life in a spectacular supernova explosion.

Why Main Sequence Stars Matter

Main sequence stars are important because they are the most stable and long-lived stars. They provide the energy that supports life on planets like Earth and are the laboratories where we learn about stellar physics.

In summary, main sequence stars are the most common type of stars, characterised by the fusion of hydrogen into helium in their cores. They come in various sizes and colours and their mass determines their lifespan and characteristics. Our Sun is a perfect example of a main sequence star and understanding these stars helps us understand the universe better.

Giants and Supernovas

Supernova: The Universe's Fireworks

Supernovas are among the most dramatic events in the cosmos. They mark the end of a star's life cycle and can be categorised mainly into two types based on their cause and characteristics: Type II and Type Ia.

- **Type II Supernova**: These occur at the end of a massive star's life when its nuclear fuel is exhausted, causing the core to collapse under gravity and resulting in a spectacular explosion.
- **Type Ia Supernova**: These are a bit different. They happen in binary star systems where one of the stars is a white dwarf—a small, dense star that's run out of fuel. The white dwarf pulls matter from its companion star and if it accumulates enough mass, it can no longer support itself and explodes. This type of supernova doesn't come from the core collapsing but from a runaway nuclear fusion reaction.

The Role of Nuclear Fusion in Type Ia Supernova

In a Type Ia supernova, the white dwarf star undergoes a thermonuclear explosion. As it gains mass from its companion, the temperature and pressure at its core increase until the carbon and oxygen present start to fuse rapidly. This fusion process releases an enormous amount of energy, leading to the white dwarf's destruction.

What makes Type Ia Supernova Special

Type Ia supernovas are particularly important in astronomy because they have a consistent peak brightness. This allows astronomers to use them as "standard candles" to measure distances in the universe. By observing how bright they appear from Earth, we can estimate how far away they are and, consequently, learn more about the expansion of the universe.

The Mystery Behind Type Ia Supernova

Despite their importance, there's still a lot we don't understand about what exactly triggers a Type Ia supernova. It's a topic of ongoing research and each discovery brings us closer to understanding the secrets of the universe.

In summary, supernovas are the grand finales of stars, with Type Ia supernovas playing a unique role in our understanding of cosmic distances and the expansion of the universe. They remind us that even in destruction, there's beauty and a wealth of knowledge to be gained.

The Solar System

The Solar System is our cosmic neighbourhood, a fascinating collection of various celestial bodies, all bound together by the gravity of the Sun.

The Sun: The Heart of the Solar System

At the centre of the Solar System is the Sun, a massive star that holds everything together with its gravity. It's a huge ball of burning gas, mostly hydrogen and makes up 99% of all the mass in the Solar System.

The Planets: A Diverse Family

Orbiting around the Sun are eight unique planets, each with its own characteristics:

- **The Terrestrial Planets**: Mercury, Venus, Earth and Mars. These are rocky planets with solid surfaces.
- **The Gas Giants**: Jupiter and Saturn. These are much larger and made mostly of gas.
- **The Ice Giants**: Uranus and Neptune. These are frozen worlds, far from the Sun.

Dwarf Planets and Other Residents

Beyond the eight planets, there are at least five recognized dwarf planets, including Pluto, Ceres and

Eris. The Solar System is also home to countless asteroids, comets and meteoroids.

Moons: Planets' Companions

Many planets and dwarf planets have moons orbiting them. These natural satellites come in all shapes and sizes, from tiny rocks to worlds with oceans and atmospheres.

The Asteroid and Kuiper Belts

Between Mars and Jupiter lies the asteroid belt, a region filled with rocky debris. Beyond Neptune, the Kuiper Belt hosts many icy bodies, including dwarf planets.

The Edge of the Solar System

The boundary of the Solar System is marked by the heliopause, where the Sun's influence ends and interstellar space begins. This is about 75–90 astronomical units from the Sun.

Why the Solar System Matters

Our Solar System is a microcosm of the universe. Studying it helps us understand how planets form and evolve and it's where we search for life beyond Earth.

Exploration and Discovery

Humanity has sent spacecraft to explore many of the Solar System's planets and moons and we continue to learn more every day. It's an ongoing adventure that expands our knowledge and inspires our imagination.

In summary, the Solar System is a dynamic place, full of wonders and mysteries. From the fiery Sun to the icy reaches of the Kuiper Belt, it's a place of incredible diversity and beauty, all moving in the grand dance of celestial mechanics.

Terrestrial vs. Gas Giants

Let's compare terrestrial planets and gas giants, the two primary types of planets in our Solar System.

Terrestrial Planets: The Rocky Worlds

Terrestrial planets are the solid, dense and rocky worlds closest to the Sun. They're made up of metals and silicate rocks and they have relatively strong gravity for their size, which allows them to hold onto an atmosphere.

- **The Inner Circle**: In our Solar System, the terrestrial planets include Mercury, Venus, Earth and Mars.
- **Surface Features**: These planets have surfaces with mountains, craters and valleys. Earth is unique with its liquid water oceans.
- **Atmospheres**: Their atmospheres are usually thin and composed mostly of carbon dioxide, nitrogen and in Earth's case, oxygen.

Gas Giants: The Massive Atmospheres

Gas giants, also known as Jovian planets, are much larger than terrestrial planets and do not have a solid surface. Instead, they are composed mainly of hydrogen and helium gases.

- **The Outer Giants**: Jupiter and Saturn are the gas giants in our Solar System.
- **Layers of Gas**: They have thick atmospheres with swirling clouds and storms. Beneath the clouds, the gases become denser and eventually turn into a liquid or metallic state.
- **Many Moons**: Gas giants have numerous moons and even their own ring systems made of ice and rock.

Key Differences

- **Composition**: Terrestrial planets are rocky, while gas giants are composed mostly of gases.
- **Size and Density**: Gas giants are larger but less dense compared to the smaller, denser terrestrial planets.
- **Location**: Terrestrial planets are closer to the Sun, whereas gas giants are found in the outer part of the Solar System.
- **Moons**: Terrestrial planets may have no moons or a few, while gas giants have many moons due to their strong gravity.

Why the Distinction Matters

Understanding the differences between terrestrial planets and gas giants helps us appreciate the diversity of worlds in our Solar System. It also aids in the search for exoplanets and the potential for life beyond Earth.

Exploring Beyond Our Solar System

When we look for planets around other stars, we find a variety of types, including gas giants and terrestrial planets. This comparison helps us categorise and understand these distant worlds.

In summary, terrestrial planets are small, rocky and dense, often with few or no moons and are located closer to the Sun. Gas giants are large, have deep atmospheres made mostly of gas and are orbited by many moons, found in the outer Solar System.

Moons

Moons, often known as natural satellites, are celestial bodies that orbit planets, dwarf planets, or even other moons. They are fascinating objects that come in various shapes, sizes and compositions. Let's explore what moons are and why they are called natural satellites.

Moons: The Travelling Companions of Planets

Moons are like the loyal companions of planets, travelling with them on their journey around the Sun. They are held in their orbits by the gravitational pull of the planet they circle.

- **Formation**: Most moons likely formed from the same cloud of gas and dust that gave birth to their parent planets. Some might have been captured by a planet's gravity and others could have formed from the debris of collisions.
- **Sizes**: Moons can be tiny, just a few kilometres across, or massive, like Jupiter's Ganymede, which is larger than the planet Mercury.

Natural Satellites: Born, Not Made

The term "natural satellite" distinguishes moons from artificial satellites, which are man-made objects sent into orbit by humans. Natural satellites are formed through natural processes in space.

- **Why 'Natural'?**: These moons weren't placed in their orbits intentionally; they either formed there or were captured by the gravity of their planet naturally.
- **Diversity**: Just like planets, moons are diverse. Some have atmospheres, some have volcanic activity and some even have subsurface oceans.

The Role of Gravity

Gravity is the force that keeps moons in orbit around their planets. It's like an invisible tether that holds them in place, preventing them from shooting off into space.

- **Stable Orbits**: For a moon to have a stable orbit, it must move at the right speed. If it's too slow, it will fall toward the planet. If it's too fast, it will escape into space.

The Importance of Moons

Moons are more than just dots in the night sky. They can affect their planets in many ways, like causing tides on Earth. Studying moons helps us understand the history and evolution of our Solar System.

Moons as Targets for Exploration

Many moons in our Solar System are targets for exploration because they might have conditions suitable for life or could tell us more about the early Solar System.

The Moon: Our Closest Neighbour

Earth's Moon is the most familiar natural satellite. It's the fifth-largest moon in the Solar System and the only other place in our Solar System where humans have set foot.

In summary, moons are natural satellites that orbit planets and other celestial bodies. They form naturally and are held in place by gravity. Moons are incredibly varied and play significant roles in shaping their planets' environments. They are also key targets for scientific research and exploration.

Black Holes

Black holes are some of the most mysterious and intriguing objects in the universe.

Black Holes: The Cosmic Enigmas

A black hole is a region in space where gravity is so strong that nothing, not even light, can escape from it. This intense gravity occurs because a lot of matter has been squeezed into a very small space.

- **Event Horizon**: The point of no return around a black hole is called the event horizon. Once something crosses this boundary, it cannot escape.
- **Singularity**: At the centre of a black hole is the singularity, where the laws of physics as we know them break down. It's a point of infinite density.

Formation of Black Holes

Black holes can form in several ways, but the most common is from the remnants of a massive star that has ended its life cycle. When such a star runs out of nuclear fuel, it collapses under its own gravity and can become a black hole.

- **Stellar Black Holes**: These are black holes that form when a massive star collapses. They can have up to 20 times the mass of the Sun.
- **Supermassive Black Holes**: These giants are millions to billions of times the mass of the Sun and are usually found at the centres of galaxies, including our own Milky Way.

Detecting Black Holes

We can't see black holes directly, but we can detect them by the effects they have on nearby matter and light. For example, if a star is orbiting a point in space but there's nothing visible there, it might be a black hole.

- **Accretion Disks**: Matter falling into a black hole can form a spinning disk around it, called an accretion disk. This matter gets very hot and can emit X-rays that we can detect.
- **Gravitational Waves**: When two black holes merge, they can produce ripples in spacetime called gravitational waves, which we can observe with special detectors.

The Mystery and Importance of Black Holes

Black holes are important for understanding the fundamental laws of the universe. They challenge our knowledge of physics, especially when it comes to gravity and the nature of spacetime.

The Future of Black Hole Research

Scientists continue to study black holes to learn more about their properties and how they affect the universe. With new tools like the Event Horizon Telescope, we're beginning to capture images of black holes and understand them better.

Stellar-Mass Black Holes: The Star's End

These black holes form when massive stars, much larger than our Sun, end their lives in a supernova explosion. What remains after the explosion is a core so dense that it collapses under its own gravity into a black hole.

- **Size**: They can be up to 20 times the mass of the Sun but are only about 10 miles across.
- **Commonality**: Stellar-mass black holes are the most common type of black holes in the universe.

Supermassive Black Holes: Galactic Giants

Supermassive black holes are the behemoths of the universe, lurking in the centres of galaxies, including our own Milky Way.

- **Size**: They can have millions to billions of times the mass of the Sun.

- **Formation**: It's still a mystery how these giants formed. They might have started as stellar-mass black holes and grown by consuming stars and gas around them or by merging with other black holes.

Intermediate-Mass Black Holes: The Missing Link

Intermediate-mass black holes are like the middle children of the black hole family. They bridge the gap between stellar-mass and supermassive black holes.

- **Size**: They are thought to range from a hundred to a few hundred thousand times the mass of the Sun.
- **Mystery**: These black holes are particularly elusive and hard to find, making them a hot topic for research.

Primordial Black Holes: Echoes of the Big Bang

Primordial black holes are hypothetical black holes that may have formed soon after the Big Bang, from the high-density fluctuations in the early universe.

- **Size**: They could range from very small, like an atom, to massive, like a mountain.
- **Significance**: If they exist, they could provide clues about the conditions of the early universe.

Why Different Types Matter

Each type of black hole gives us unique insights into the workings of the universe. By studying them, we can learn about the life cycles of stars, the formation of galaxies and the evolution of the cosmos.

The Ongoing Mystery

Black holes remain one of the greatest mysteries in astrophysics. With new observational technologies, we continue to uncover more about these enigmatic objects and their role in the universe.

In summary, black holes come in various types, primarily categorised as stellar-mass, supermassive, intermediate-mass and theoretical primordial black holes. Each type has its own characteristics and origins, contributing to our understanding of the universe's most mysterious phenomena.

Wormholes

Wormholes are one of the most captivating concepts in astrophysics, often featured in science fiction as shortcuts through space and time.

Wormholes: The Universe's Tunnels

Imagine if the universe had secret tunnels that could connect two very distant places, making travel between them much shorter than going the long way around. That's what a wormhole is like—a hypothetical shortcut through the fabric of space and time.

- **The Basics**: A wormhole can be visualised as a tunnel with two ends, each at separate points in spacetime. This means they could potentially connect different locations, times, or even different universes.
- **Einstein's Theory**: Wormholes come from solutions to Einstein's equations of general relativity, which describe how gravity works. They are consistent with the theory, but we don't know if they actually exist.

Visualising Wormholes

To help visualise a wormhole, think of space as a two-dimensional sheet. If you fold the sheet so two points touch, a wormhole could be the bridge connecting those points.

- **Entry and Exit**: The ends of a wormhole are called 'mouths,' and the passage between them is the 'throat.' Theoretically, you could enter one mouth and exit the other almost instantaneously.

Types of Wormholes

There are different types of wormholes theorised:

- **Schwarzschild Wormholes**: These are also known as Einstein-Rosen bridges and they connect two black holes. However, they're not stable and would collapse too quickly for anything to cross.
- **Traversable Wormholes**: These are the kind that you might see in movies, where you can actually travel from one end to the other. They would require some form of 'exotic matter' with negative energy to keep them open.

The Challenges

Wormholes face several scientific challenges:

- **Stability**: For a wormhole to be traversable, it needs to be stable. This means preventing it from collapsing on itself, which would require matter that behaves in ways we've never observed.

- **Creation**: We don't know how to create a wormhole. Some scientists have proposed methods, but they remain theoretical.

The Role of Wormholes in Science Fiction

In science fiction, wormholes are often used as a plot device to allow characters to travel vast distances quickly or even time travel. They capture our imagination with the possibility of exploring the far reaches of the universe or different timelines.

The Future of Wormhole Research

While wormholes are still theoretical, research continues. Scientists are looking for signs that could indicate the presence of wormholes and new theories and technologies might one day make the concept of wormholes a reality.

In summary, wormholes are theoretical tunnels in spacetime that could connect distant points in the universe. They arise from Einstein's theory of general relativity and while they are a staple of science fiction, their existence in reality is still unconfirmed. The study of wormholes challenges our understanding of physics and fuels our dreams of distant space travel.

Types of Galaxies

Galaxies are vast islands of stars, gas and dust. They come in various shapes and sizes. Let's explore the three main types of galaxies: Spiral, Elliptical and Irregular.

Spiral Galaxies: The Cosmic Pinwheels

Spiral galaxies are like grand cosmic pinwheels, with arms winding out from a central bulge. They're flat and disk-shaped with older stars at the centre and younger stars along the arms.

- **Features**: They have a central bulge surrounded by a flat, rotating disk of stars and spiral arms.
- **Star Formation**: The arms of spiral galaxies are often sites of active star formation, giving them a bluish tint.
- **The Milky Way**: Our own galaxy, the Milky Way, is a spiral galaxy.

Elliptical Galaxies: The Star-Studded Spheres

Elliptical galaxies range from nearly spherical to slightly elongated and lack the distinct arms of spiral galaxies. They're made up mostly of older, redder stars.

- **Structure**: They have a smooth, featureless light distribution and can range from nearly spherical to slightly elongated shapes.
- **Size**: Elliptical galaxies vary greatly in size and can be massive, containing trillions of stars, or quite small.
- **Star Content**: They generally contain older stars and have little gas or dust, which means less new star formation.

Irregular Galaxies: The Unconventional Outliers

Irregular galaxies don't fit into the spiral or elliptical categories. They have an irregular shape and often appear chaotic, without a central bulge or spiral arms.

- **Characteristics**: They are often rich in gas and dust, with regions of intense star formation.
- **Examples**: The Large and Small Magellanic Clouds, which are neighbours to the Milky Way, are irregular galaxies.

Why Galaxy Types Matter

Understanding the different types of galaxies helps astronomers study the universe's evolution. Each type holds clues about the conditions in the early universe and the processes that shape galaxies.

Exploring the Universe

Astronomers use telescopes to observe these galaxies across vast distances, effectively looking back in time to understand how galaxies form and evolve.

In summary, galaxies are classified into three main types based on their shape: Spiral galaxies with their distinctive arms, elliptical galaxies with their older star populations and irregular galaxies with their chaotic appearances. Each type provides unique insights into the workings of the cosmos and the history of the universe.

Galaxy Clusters

Galaxy clusters are some of the largest and most massive structures in the cosmos, fascinating for their grand scale and the role they play in the universe's structure. Let's explore galaxy clusters!

Galaxy Clusters: The Universe's Great Cities

Think of galaxy clusters as the universe's great cities, where hundreds to thousands of galaxies live together, bound by gravity. They're like metropolises filled with diverse galaxy inhabitants.

- **Massive Structures**: Galaxy clusters can have masses ranging from 10^{14} to 10^{15} times the mass of our Sun. That's incredibly heavy
- **Components**: They contain not just galaxies, but also hot gas emitting X-rays and a lot of dark matter, which we can't see but know is there because of its gravitational effects.

The Intracluster Medium: The Space Between

The space between the galaxies in a cluster isn't empty. It's filled with an intracluster medium (ICM) of heated gas. This gas is so hot that it shines in X-ray light.

- **Temperature**: The ICM can reach temperatures between 2–15 keV, depending on the cluster's total mass.
- **X-rays**: Because the gas is so hot, it emits X-rays, which astronomers can observe with special telescopes.

Gravitational Giants

Galaxy clusters are held together by gravity. They're so massive that they can even warp the fabric of spacetime, creating phenomena like gravitational lensing, where the light from objects behind the cluster is bent.

Galaxy Groups vs. Clusters

Smaller collections of galaxies, typically with fewer than 50 members, are called galaxy groups. Our own Milky Way is part of the Local Group, which is like a small town compared to the big cities of galaxy clusters.

Superclusters: Clusters of Clusters

Galaxy clusters can come together to form even larger structures called superclusters. These are not as tightly bound by gravity but are still impressive gatherings of galaxy clusters.

Studying Galaxy Clusters

Astronomers study galaxy clusters to learn about the large-scale structure of the universe and the processes that shape galaxies. Clusters can tell us about the way galaxies form and evolve over time.

The Role of Dark Matter

Dark matter plays a crucial role in the formation and structure of galaxy clusters. It provides the extra gravity needed to hold these massive structures together.

In summary, galaxy clusters are the largest gravitationally bound structures in the universe, containing hundreds to thousands of galaxies, hot gas and dark matter. They serve as "cities" in the cosmos, providing homes for galaxies and offering insights into the universe's evolution and the nature of dark matter.

Dark Matter

Dark matter is one of the most intriguing and elusive components of the universe. Let's delve into what dark matter is, its importance and the mysteries surrounding it.

Dark Matter: The Invisible Substance

Dark matter is a form of matter that does not emit, absorb, or reflect light, making it invisible to current telescopic technology. Its presence is known only because of the gravitational effects it has on visible matter, radiation and the large-scale structure of the universe.

- **Gravitational Effects**: We can detect dark matter because of the way it influences the orbits of stars in galaxies and the movement of galaxies within clusters.
- **The Cosmic Recipe**: Dark matter makes up about 26.8% of the universe's mass-energy composition, while dark energy comprises 68.2% and visible matter makes up just 5%.

Evidence for Dark Matter

The existence of dark matter is inferred from several astronomical observations:

- **Galactic Rotation Curves**: The rotation speeds of stars in galaxies remain constant at distances where they should decrease according to the visible mass, suggesting the presence of additional, unseen mass.
- **Gravitational Lensing**: The bending of light from distant objects by galaxy clusters can't be explained by the visible mass alone, indicating extra mass provided by dark matter.
- **Cosmic Microwave Background**: The temperature fluctuations in the cosmic microwave background radiation also suggest the presence of dark matter influencing the early universe's structure.

What Could Dark Matter Be?

The true nature of dark matter is still unknown, but there are several candidates:

- **WIMPs**: Weakly Interacting Massive Particles are hypothetical particles that rarely interact with ordinary matter except through gravity.
- **Axions**: Another theoretical particle that is light and interacts very weakly with ordinary matter.
- **Primordial Black Holes**: Some theories suggest dark matter could be made of black holes formed in the early universe.

Searching for Dark Matter

Scientists are actively searching for dark matter through various experiments and observations:

- **Direct Detection**: Experiments deep underground aim to detect dark matter particles as they pass through the Earth.
- **Indirect Detection**: Observatories look for signs of dark matter particles annihilating or decaying into standard particles in space.

The Mystery Continues

Despite decades of research, dark matter has yet to be directly detected and it's true nature remains one of the biggest mysteries in physics.

The Accelerating Universe: A Speeding Cosmos

The universe is not just expanding; it's expanding at an increasing rate. This surprising discovery was made in 1998 when astronomers observed distant supernovas and found that they were moving away from us faster than expected.

- **Dark Energy**: The force thought to be responsible for this acceleration is called dark

energy. It's a mysterious form of energy that seems to be pushing galaxies apart, overcoming the gravitational pull that should be slowing them down.
- **Cosmological Constant**: One explanation for dark energy is the cosmological constant, a term in Einstein's equations of general relativity that represents a constant energy density filling space.

The Relationship Between Dark Energy and Dark Matter

While dark energy is driving the expansion of the universe to accelerate, dark matter is working to slow it down by gravitationally attracting matter. Together, they make up about 95% of the total mass-energy content of the universe, with dark energy being the dominant force in the current era.

- **Cosmic Balance**: The interplay between dark energy and dark matter determines the fate of the universe. If dark energy continues to dominate, the universe may expand forever. If dark matter were to somehow prevail, the expansion could reverse, leading to a "big crunch".

Unravelling the Mysteries

Both dark energy and dark matter are subjects of intense study. Understanding them is crucial for a complete picture of the universe's past, present and future.

In summary, the accelerating universe is a phenomenon driven by dark energy, which causes the expansion of space to speed up. Dark matter, while not related to this acceleration, is another unseen component that exerts a gravitational pull on galaxies. Together, these dark components shape the structure and destiny of the universe.

Why Dark Matter Matters

Understanding dark matter is crucial for our comprehension of the universe. It plays a key role in the formation and evolution of galaxies and the overall structure of the cosmos.

In summary, dark matter is an invisible component of the universe, detectable only through its gravitational effects. It is a fundamental part of the universe's structure and uncovering its secrets could revolutionise our understanding of physics and the cosmos.

Exoplanets

Exoplanets, also known as extrasolar planets, are planets that orbit stars outside our own Solar System. They are a diverse group of celestial bodies that have captured the interest of astronomers and the public alike.

Understanding Exoplanets

Exoplanets are planets that do not orbit the Sun but other stars, much like our own planets orbit our star, the Sun.

- **Discovery**: The first confirmed detection of an exoplanet occurred in 1992.
- **Numbers**: As of April 2024, there are 5,653 confirmed exoplanets in 4,161 planetary systems, with some systems having multiple planets.

Methods of Detection

There are several methods used to discover exoplanets:

- **Transit Photometry**: This method detects the dimming of a star's light when a planet passes in front of it.
- **Doppler Spectroscopy**: Also known as the radial velocity method, it measures changes in a

star's spectrum caused by the gravitational pull of an orbiting planet.

Types of Exoplanets

Exoplanets come in various types, much like the planets in our Solar System:

- **Gas Giants**: These are large planets, similar to Jupiter and Saturn, which are mostly composed of hydrogen and helium.
- **Rocky Planets**: Smaller, terrestrial-like planets with solid surfaces, akin to Earth or Mars.
- **Ice Giants**: Planets that have a solid core surrounded by a thick layer of gas and ice, like Uranus and Neptune.

Habitable Zones and Life

The habitable zone, often referred to as the "Goldilocks Zone," is the region around a star where conditions might be just right for liquid water to exist on a planet's surface—a key ingredient for life as we know it.

- **Potential for Life**: Planets in the habitable zone are of particular interest because they have the potential to harbour life.

The Future of Exoplanet Research

With advancements in technology, such as the James Webb Space Telescope, astronomers hope to learn more about the composition, environmental conditions and potential for life on exoplanets.

Why Exoplanets Fascinate Us

The study of exoplanets helps us understand the formation and evolution of planetary systems and the possibility of finding life beyond Earth. It expands our knowledge of the universe and our place within it.

In summary, exoplanets are planets that orbit stars other than the Sun. They are discovered through various methods and come in many forms, from gas giants to rocky planets. The study of exoplanets is a rapidly growing field that promises to reveal new insights into the cosmos and the potential for life beyond our Solar System.

Copyright © 2024 by For Noobs
All rights reserved

www.ingramcontent.com/pod-product-compliance
Lightning Source LLC
Chambersburg PA
CBHW030051230526
45471CB00003B/1046